I0462339

Introduction
to
GSM

Second Edition

by

Guy Inchbald

steelpillow   Lulu

First Edition published 2005
by Guy Inchbald

Second Edition 2017
Reprinted 2021
published by steelpillow
in association with Lulu.com

Copyright © Guy Inchbald 2005, 2017, 2021

All rights reserved. This book or any portion thereof may not
be reproduced, downloaded or used in any manner whatsoever
without the express written permission of the publisher, except
for the fair use of brief quotations in a book review or
scholarly discussion.

Set in Palladio 10.5 pt font.

steelpillow, Park View, Queenhill,
Upton-on-Severn, Worcs. WR8 0RE, UK

ISBN   978-0-244-91368-7

Contents

Preface to the Second Edition 7

1 Cellular mobile telephony 8

1.1 Origins 8

1.2 The GSM system 9
1.2.1 Circuit switching 9
1.2.2 CCS – Common Channel Signalling 9
1.2.3 Evolution 10

2 The GSM network 11

2.1 The cellular system 11
2.1.1 Frequency re-use and clustering 11
2.1.2 Types of cell 13
2.1.3 Physical architecture 14
2.1.4 Geographical areas 15

2.2 MS – Mobile Station 15
2.2.1 ME – Mobile Equipment 16
2.2.2 SIM – Subscriber Identity Module 16

2.3 BSS – Base Station Subsystem 17
2.3.1 BTS – Base Transceiver Station 17
2.3.2 BSC – Base Station Controller 17

2.4 NSS – Network and Switching Subsystem 17
2.4.1 MSC – Mobile services Switching Centre 18
2.4.2 HLR – Home Location Register 18
2.4.3 VLR – Visitor Location Register 18
2.4.4 AuC – Authentication Centre 19
2.4.5 EIR – Equipment Identity Register 19
2.4.6 GIWU – GSM Interworking Unit 19

2.5 OSS – Operation and Support Subsystem 19

3 Network functions and signalling 20

3.1 GSM network functions 20

3.2 Transmission 20

3.3 RR – Radio Resources management 20
3.3.1 Handover 21

3.4 MM – Mobility Management 22
3.4.1 Location management 22
3.4.2 Authentication and security 23

3.5 CM – Communication Management 24
3.5.1 CC – Call Control 24
3.5.2 Supplementary Services management 24
3.5.3 SMS (Short Message Services) management 25

3.6 OAM – Operation, Administration and Maintenance 25

3.7 Signalling architecture 26
3.7.1 Layer 3 protocols 26
3.7.2 Layer 2 – LAPDm 26
3.7.3 Layer 1 27

4 The air interface 28

4.1 The GSM radio bands 28

4.2 Multiple access scheme 29
4.2.1 FDMA 29
4.2.2 TDMA 30

4.3 Other radio features 31
4.3.1 Timing advance 31
4.3.2 Power control 31
4.3.3 Multipath equalisation and the training sequence 32
4.3.4 Frequency hopping 32
4.3.5 DTX – Discontinuous transmission 33
4.3.6 Discontinuous reception 33

4.4 Processing the signal from source to destination 34
4.4.1 Data coding 34
4.4.2 Channel coding 35
4.4.3 Interleaving 36

4.4.4 Burst assembling 36
4.4.5 Enciphering 36
4.4.6 Modulation 37

5   Channel structures **39**

5.1   General structures 39
5.1.1 Logical channels and grouping 39
5.1.2 TDMA frame sequencing 40
5.1.3 Burst structure 41

5.2   User traffic and associated signalling 42
5.2.1 TCH – Traffic channels 42
5.2.2 SACCH – Slow Associated Control Channels 43
5.2.3 FACCH – Fast Associated Control Channels 43
5.2.4 The 26-frame multiframe 43

5.3   Common access control 44
5.3.1 BCH – Broadcast Channels 44
5.3.2 CCCH – Common Control Channels 46

5.4   Dedicated signalling 46
5.4.1 DCCH – Dedicated Control Channels 46

6   GSM services **48**

6.1   Teleservices 48

6.2   Bearer services 49

6.3   Supplementary services 49

List of acronyms **51**

Figures

Fig	1	Cells and clustering	12
Fig	2	GSM network architecture	14
Fig	3	Geographical areas in GSM	15
Fig	4	Authentication	23
Fig	5	The signalling network model	26
Fig	6	Time-division multiplexing	30
Fig	7	Processing of speech to radio and back	35
Fig	8	Enciphering	37
Fig	9	The TDMA frame and normal burst	41
Fig	10	Type III 26-multiframe	44

Preface to the Second Edition

This brief introduction to GSM "2G" cellular radio telephony is intended both as an introductory guide to the new student and as a handy *aide memoire* for those working in the field.

A technical overview introduces the principles of mobile radio telephony in general. After outlining basic concepts and the GSM high-level architecture, the challenges of mobile telephony are emphasised through the complexities of the air interface. Finally, the services carried by the GSM system are briefly identified.

GSM jargon can be confusing, not only due to the many acronyms, but also because some technical terms, for example "channel" and "frame", can have different meanings in different contexts. I have tried to keep things simple. A list of acronyms is provided at the end.

Major changes introduced for this edition include a handier A5 format and an updated and expanded first chapter. There are many small corrections and updates, with more included in this reprint.

Since I first wrote it over ten years ago, the features built onto the basic GSM system evolved steadily. Chief among these has been the rise and fall of "2½G" (or 2.5G) GPRS. Many of the backhaul and management subsystems have been progressively virtualised or otherwise embedded into modern hybrid hosts. I have skimmed very briefly over such complications: the original system provides an excellent demonstration of the basic principles and its functions must still be supported.

If you have any comments, especially if you spot any mistakes or miss some useful nugget, I would be glad to hear from you.

Guy Inchbald, MA, BA, B.Sc, FRSA
guy@steelpillow.com

June 2017, March 2021

1 Cellular mobile telephony

1.1 Origins

Any two-way radio communication device needs a *transmitter* to send signals and a *receiver* to pick up the return signal. The signals travel through the air as waves. Such a wave is characterised by its *frequency* (expressed as cycles per second or Hertz) and strength or *amplitude* (don't ask). Different frequencies may be used to send out different signals, even from the same transmitter. The range of frequencies which the transmitter is able to use is called its frequency *band*. A transmitter needs to send out a strong signal and a receiver to pick up a weak one, because most of the wave will dissipate elsewhere.

Telephony also needs a sender and receiver at each end of the phone call. The signal is then switched through the provider's network to create an end-to-end connection.

A cellular radio telephony system has many radio base stations, each comprising one or more transmitter and receiver pairs. Each handset user connects to the nearest base station. The base stations then connect through to the provider's network. The user may roam from one base station's catchment area, or cell, to another – sometimes during a call. This is the fundamental feature which takes mobile telephony beyond a mere radio link.

The first such systems were developed in the early 1970s using analogue technology, with the first generation (1G) commercial services introduced in the 1980s. The emerging digital technology soon offered huge advances in performance, flexibility, cost and compatibility with other new digital telephony systems such as ISDN.

1.2 The GSM system

The Global System for Mobile communications (GSM) was the first second-generation (2G), i.e. digital, cellular system and is still very widely deployed. Developed as a common European standard, it was rapidly accepted worldwide. The Conference of European Posts and Telecommunications (CEPT) set up the Groupe Spécial Mobile (GSM) in 1982. The GSM acronym later changed to its current meaning, and responsibility passed to the European Telecommunications Standards Institute (ETSI). Phase 1 of the GSM specifications was published in 1990 and Phase 2, on coverage of rural areas, in 1995.

Besides the general benefits of going digital, GSM added international roaming; you could use your cellphone in other countries which had adopted the same standard. Technically, GSM is a circuit switched digital cellular mobile radio communications system.

1.2.1 Circuit switching

GSM still uses the old-fashioned idea of connecting a call through by setting up a dedicated circuit or channel across the network. Digital techniques allow for such circuits to be ephemeral. They are switched through for each call and torn down again afterwards. This generates a great deal of signalling to manage the channels. It represents an advance on the old hardwired circuits and in-band dialling of the analogue landline days, while it is still very different from the data oriented packet-switching system used for the Internet.

1.2.2 CCS – Common Channel Signalling

One advance made popular by going digital is worth special note. Originally, call setup and teardown information was passed in the same communications channel as the call audio. In Common Channel Signalling, a special *signalling channel* is used to carry the signalling data for many calls. The call channels then only carry the actual voice or data content of the call.

GSM is such a CCS system. In order to maintain the communication links within the cellular network, several radio channels may be reserved for the various kinds of signalling information.

Engineers familiar with the land-based System 7 signalling (SS7, C7, etc) will find many familiar acronyms with an 'm' suffix added – LAPDm, Um, and so on – the m indicates adaptation for mobile use.

1.2.3 Evolution

Packet-switched Internet traffic is unsuited to a circuit-switched system tailored for voice telephony. A stopgap technology called GPRS was grafted onto the basic GSM infrastructure to create a so-called 2½G (or 2.5G) system. The physical radio interface was not affected, but the logical data and channel structures were entirely different. Later enhancements to GPRS included EDGE and HSCSD.

But the data rates obtainable from 2½G were being outstripped by demand even as it was being rolled out, so more advanced 3G and later 4G systems appeared, bringing not only faster data links but a level of convergence between voice and Internet traffic systems.

However they in their turn were overly data-oriented and became sluggish when overburdened with voice calls, crippling the hoped-for advances in the Mobile Internet. Network providers have found it increasingly convenient to put up hybrid 2G/4G and similar links where the voice goes one way, onto a GSM link designed for it, and the Internet goes another. The extra ½G tacked onto GSM has no sensible function in such a design and is beginning to wither.

Meanwhile many of the backhaul and management GSM subsystems have become virtualised or otherwise adapted to an assortment of new hybrid carrier systems. The original system remains the GSM springboard for all those later complications and they must all remain compatible at a functional level. Even though some GSM networks have been wholly replaced by more modern ones, the basic 2G GSM system design is likely to have a long life ahead of it.

2 The GSM network

2.1 The cellular system

In a cellular system, the geographical covering area of a network operator is divided into many *cells*. A cell corresponds to the local covering area of one transmitter or *base station*. The size of a cell is directly related to the transmitter's power. A band of radio frequencies is allocated to GSM, with each cell operating on just a few of the available frequencies within the band.

A cellular system must meet two main conditions:

- Neighbouring cells can not operate on the same radio frequency. The distance between cells using the same frequency must be sufficient to avoid interference.
- The power level of a transmitter within a single cell must be limited in order to reduce interference with nearby cells using the same frequency. Generally a distance between transmitters of about 2.5 to 3 times the diameter of a cell is required.

2.1.1 Frequency re-use and clustering

The set of frequencies allocated to a given system is distributed over a group, or *cluster*, of cells. All the cells in the cluster operate on different frequencies so that they do not interfere with each other. The same set of frequencies is then re-used in adjacent clusters, with the same basic cluster pattern repeated many times over the covering area of the network.

Fig 1 shows an example in which seven cells form a cluster (within the thick black outline). Each cell in the cluster operates on a different set of frequencies, here numbered 1 to 8. The cluster on the right is repeated over on the left, and partially again below. Cells sharing a common frequency, such as those marked 1, are far enough apart not to interfere with each other. This frequency re-use by many clusters hugely increases the call capacity of the network.

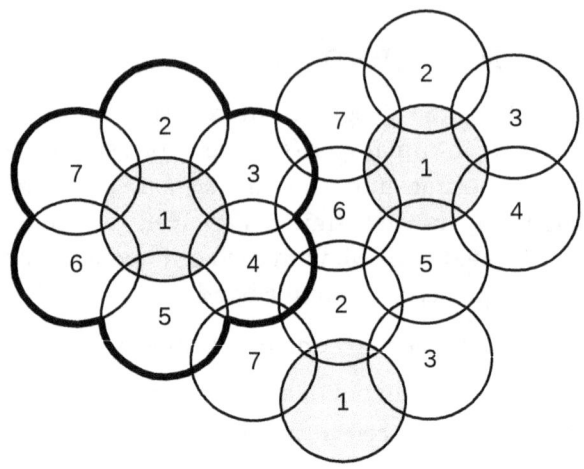

Fig 1 : Cells and clustering

The GSM band provides for 124 radio channel frequencies. There are typically 4, 7, 12 or 21 cells in a cluster. The smaller the cluster, the more frequencies are available to each cell, increasing the capacity of the cell. However if the clusters are too small, the cells using a given frequency will be physically too close together, despite being in different clusters, and will interfere. The network designer must balance these conflicting demands. A cell may typically be allocated from 1 to 16 frequencies.

2.1.2 Types of cell

Different types of cell are used according to the characteristics of the local mobile population. Specialised types include:

- **Hypercells or Macrocells.** Hypercells are large cells up to 20 km radius, and are used for remote and sparsely populated areas. They can transmit at high power and must have sensitive receivers. The reduced number of cells leads to lower overheads.

- **Microcells.** Microcells have a radius from 50 to 300 m and are used for densely populated areas. By splitting the area into many smaller cells, the total number of channels available within the area is increased. The transmitter power used is quite small, to reduce the risk of interference between nearby cells.

- **Selective or directional cells.** It is not always desirable for a cell to have full 360 degree coverage. In some cases selective, or directional, cells having a particular shape and coverage are needed. A typical example is the entrance to a tunnel, where a coverage of say 60 degrees may be used. A more sophisticated example is the **smart cell**, which dynamically changes its coverage as required, comprising for example a pencil-beam and directional receiver which can follow a small group of mobiles.

- **Umbrella cells.** A busy transport link crossing many small cells would produce frequent handovers between cells as travellers whizz by, creating excessive signalling and switching load on the network. To reduce this load and improve reliability, fast-moving mobiles are handed over to an umbrella cell, extending across several microcells. The power level of the umbrella cell is higher than the microcells, and of course care must be taken that the microcells leave some frequencies free for the umbrella to use.

The speed of a mobile can be deduced either from its pattern of handover demands or from its radio propagation characteristics.

2.1.3 Physical architecture

The physical components of the GSM network are defined according to their function and interface requirements. The four main components, their general relationships and internal arrangements are shown in Fig 2.

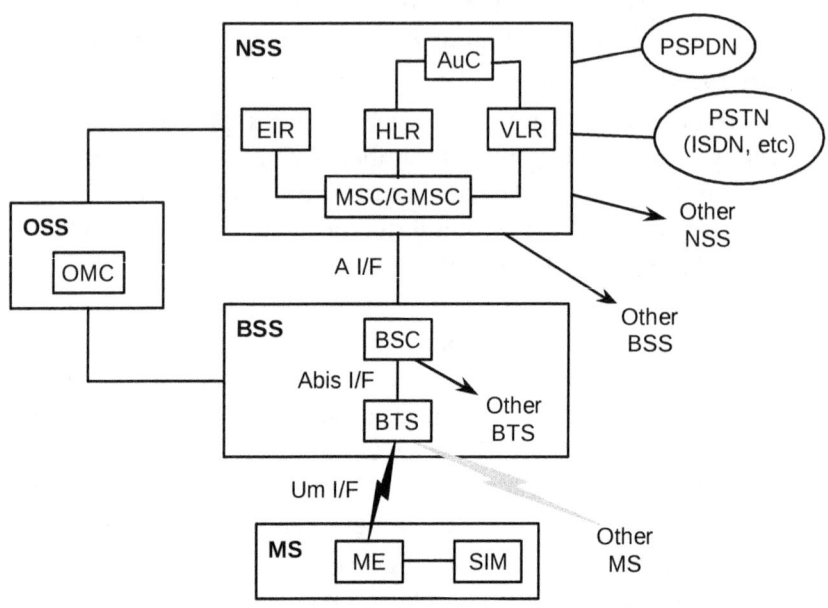

Fig 2 : GSM network architecture

The MS (mobile station) is the familiar mobile phone, and the BTS (base transceiver station) the fixed base station for a cell. The NSS is concerned with network switching and security, and the OSS with general support and maintenance.

2.1.4 Geographical areas

An operator's GSM network comprises a hierarchy of five types of area, as shown in Fig 3.

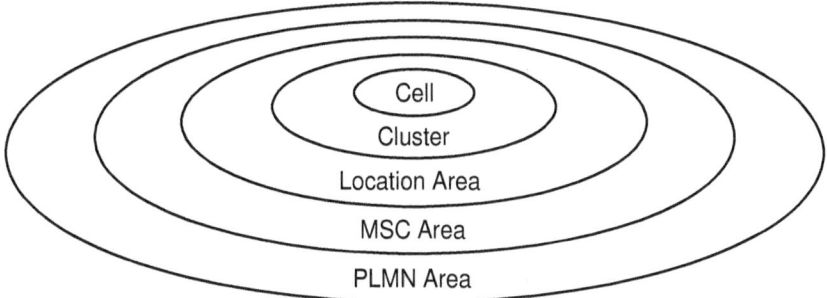

Fig 3 : Geographical areas in GSM

- **Cell.** The basic unit of the cellular system, being the radio coverage area of a single Base Transceiver Station. Identified by its Cell Global Identity number (CGI).
- **Cluster.** A group of cells having no common frequency, and repeated over the covering area. Has no special ID.
- **LA – Location Area.** A logical subgroup of cells served by the same switching centre. Identified by its Location Area Identity (LAI) number.
- **MSC area. The** whole area controlled by a given switching centre (MSC or GMSC).
- **PLMN area. The** whole geographic covering area served by the operator's Public Land Mobile Network.

2.2 MS – Mobile Station

A Mobile Station consists of two main elements:

- Mobile Equipment (ME).
- Subscriber Identity Module (SIM).

2.2.1 ME – Mobile Equipment

Three types of mobile equipment are defined, mainly by their power and typical use:

- **Handheld terminals** are by far the most common. Their maximum allowed output power is 2W, though this may be reduced to 0.8 W as technologies develop.
- **Portable terminals** may be installed in vehicles. Maximum allowed power is 8W.
- **Fixed terminals** are usually installed in vehicles. Maximum allowed power is 20W.

Most ME are identified by a unique International Mobile Equipment Identifier (IMEI) code, to aid in compatibility and security management.

2.2.2 SIM – Subscriber Identity Module

The SIM is a smart card that identifies the subscriber. In fact, it is the only element that personalises a terminal to the GSM system. The user can have access to their subscribed services from any compatible terminal using their SIM. Without a SIM, the mobile equipment cannot be used (except for 999 calls).

The SIM is protected by a four-digit Personal Identification Number (PIN) and, in case the PIN is forgotten, an 8-digit Personal Unlocking Key (PUK).

In order to identify the subscriber to the system, the SIM contains other identifying parameters such as the International Mobile Subscriber Identity (IMSI) and the A3 algorithm used for enciphering the IMSI.

The SIM may also contain non-GSM personal information and functionality, such as stored names and phone numbers.

SIM card data tends to be heavily protected against tampering, including self-destruction on any attempt at hacking or at physical cutting-open and microscopic examination.

2.3 BSS – Base Station Subsystem

The BSS connects the Mobile Station to the NSS. It is in charge of transmission and reception. The BSS can be divided into two parts:

* The Base Transceiver Station (BTS) or Base Station.
* The Base Station Controller (BSC).

2.3.1 BTS – Base Transceiver Station

The BTS provides the transceivers and antennas used in each cell of the network. A BTS is usually placed in the centre of its cell. Its transmitting power defines the size of the cell. Each BTS has between one and sixteen transceivers depending on the density of users in the cell, with each transceiver operating on a given frequency channel.

2.3.2 BSC – Base Station Controller

The BSC controls a group of BTS (from tens to hundreds) and manages their radio resources. It is principally in charge of hand-overs, frequency hopping, exchange functions and control of the transmitter power levels of the BTSs.

2.4 NSS – Network and Switching Subsystem

The main role of the NSS is to manage the communications between the mobile users and other users, such as mobile users, ISDN users, fixed telephony users, etc. It includes databases which store information about the subscribers required to manage their mobility. The NSS has several sub-components.

2.4.1 MSC – Mobile services Switching Centre

The MSC is the heart of the NSS. It performs the main datalink switching functions of the network, connecting between the various BSS and neighbouring MSCs.

Where an NSS acts as a gateway to the public PSTN, a GMSC – Gateway Mobile services Switching Centre – is used. This has the added capacity to route calls between mobile users and the PSTN network (e.g. ISDN). A GMSC requires an associated database of subscriber information for its covering area, the HLR.

This booklet uses the term MSC/GMSC to indicate that either may be involved.

2.4.2 HLR – Home Location Register

The HLR is a database which stores subscriber information for the covering area of the GMSC, including the services to which they have access. It also records the current VLR (see below) registering the subscriber.

Permanent information held by the HLR includes the user's IMSI, PSTN, authentication key, and details of subscribed services.

Temporary information includes the current VLR (in the form of its SS7 address), transient authentication and ciphering keys, and any forwarding number.

A BSS which is not connected to the PSTN, having just an MSC, does not need an HLR.

2.4.3 VLR – Visitor Location Register

The VLR contains information about visiting users, copied from the subscriber's HLR. When a user enters the covering area of a new MSC/GMSC, the associated VLR will request information about the subscriber from their corresponding HLR. The VLR can then provide the subscribed services without constant accesses to the HLR.

The VLR is always accompanied by a MSC/GMSC, so the area under control of the MSC/GMSC is also the area under control of the VLR, and is sometimes referred to as the MSC/VLR area.

2.4.4   AuC – Authentication Centre

The AuC supports the two security functions:

* Authentication of the subscriber.
* Enciphering, or encryption, of user data.

The AuC maintains a database of the parameters needed for these functions, and provides associated data processing power.

2.4.5   EIR – Equipment Identity Register

The EIR also supports authentication. It contains a list of all valid mobile equipments, identified by their IMEI. The EIR will bar any calls from stolen or unauthorised terminals (for example one which breaches the rules for RF output power).

Older GSM networks may not have an EIR, and not all terminals have an IMEI.

2.4.6   GIWU – GSM Interworking Unit

The GIWU provides the interface to other communications networks, for the transmission of speech and/or data.

2.5   OSS – Operation and Support Subsystem

The OSS comprises the Operation and Maintenance Centre (OMC). It links to the different components of the NSS and to the BSC, in order to control and monitor the GSM system.

The OSS also controls the traffic load of the BSS.

3 Network functions and signalling

3.1 GSM network functions

In GSM, five main network system functions are defined:

- Transmission.
- Radio Resources management (RR).
- Mobility Management (MM).
- Communication Management (CM).
- Operation, Administration and Maintenance (OAM).

The three management functions – RR, MM and CM – generate all the signalling on the air interface to and from the mobile.

3.2 Transmission

The transmission function includes two sub-functions, namely the transmission of user information (traffic) and of signalling information.

The MS, the BTS and the BSC are the main components concerned with transmission. Section 4 describes the transmission functions in more detail.

3.3 RR – Radio Resources management

The RR network function establishes, maintains and releases communication links between mobile stations and the MSC. The elements mainly concerned with the RR function are the mobile station and the base station. The RR function must maintain a connection even when the user moves from one cell to another, so the MSC is also involved.

The RR function also manages the frequency spectrum and the reaction of the network to changing radio environment conditions.

Some of the main RR procedures are:

◆ Channel assignment, change and release.

◆ Handover.

◆ Frequency hopping.

◆ Power-level control.

◆ Discontinuous transmission and reception.

◆ Timing advance.

Some of these are visited in section 4. In this paragraph only the handover, one of the most important responsibilities of the RR, is described.

3.3.1 Handover

The mobile continuously monitors both its own signal strength and that of the neighbouring cells. The base station tells it which cells to monitor, and then uses its power measurements to decide which cell can best maintain the quality of the communication link. If a change is needed, the mobile must be moved to a different channel or cell. This is called handover. It can occur in four situations:

◆ Between channels in the same cell.

◆ Between cells controlled by the same BSC.

◆ Between cells belonging to the same MSC but controlled by different BSCs.

◆ Between cells controlled by different MSCs.

The first two types of handover are managed by the local BSC (and the MSC is notified of the handover). The second two types are controlled by the MSC.

Two basic algorithms are used for handover:

Power budget. Where it is not necessary to increase the power level in order to obtain a good communication quality, this algorithm performs the handover.

Minimum acceptable performance. Where the transmission quality to the new base station is poor, the power level of the mobile is increased until any further increase has no effect on the quality of the signal. The handover is then carried out.

3.4 MM – Mobility Management

The MM function oversees all other aspects related to the mobility of the user, especially location management and security authentication.

The MM functions need to identify the subscriber. This is done by allocating a unique IMSI number which is held in both the mobile's SIM and in the subscriber's HLR. The IMSI is structured as follows:

1st 3 digits	MCC – Mobile Country Code (e.g. the UK is 234).
Next 2 digits	NCC – Network Country Code (e.g. UK Cellnet (O_2) is 10).
Up to 10 digits	MSIC – Mobile Subscriber Identification Code.

3.4.1 Location management

When a mobile station is powered on, it performs a location update procedure by passing its IMSI to the network. This identifies its current location. This first location update procedure is called the *IMSI attach* procedure.

The mobile station performs a location update whenever it moves to a new Location Area or a different PLMN. The location update is passed to the new MSC/VLR, which sends the new location information to the subscriber's HLR. If the mobile station is authorised with the new MSC/VLR, the subscriber's HLR cancels their registration with the previous MSC/VLR.

Periodic location updates are also used to check that the mobile is still active. If the mobile station has not updated when due, it is de-registered.

When a mobile station is powered off, it performs an IMSI detach procedure to tell the network that it is no longer connected.

3.4.2 Authentication and security

A user authentication procedure is carried out between the SIM card and the Authentication Centre. A secret key Ki, stored in the SIM card and in the AuC, a random number (RAND) generated by the AuC, and an enciphering algorithm called A3 are used. From them the mobile station and the AuC each compute a 32-bit Signed RESult (SRES). If the two computed SRES are the same, the subscriber is authenticated. Note that only the RAND and SRES are transmitted, the key Ki remains private.

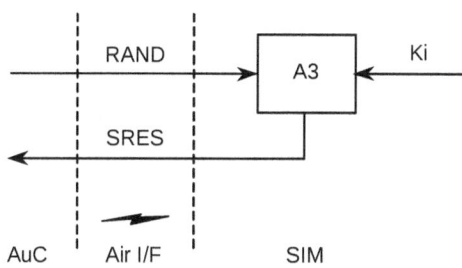

Fig 4 : Authentication

The different services to which the subscriber has access are checked. The mobile equipment identity may also be checked – if its IMEI number is not authorised in the EIR, it may not be allowed to connect to the network.

To assure user confidentiality, the user is registered with a Temporary Mobile Subscriber Identity (TMSI) after the IMSI attach procedure.

Enciphering is another security option – see section 4.4.5.

3.5 CM – Communication Management

The CM function is responsible for:

- Call control.
- Supplementary Services management.
- SMS (Short Message Services) management.

3.5.1 CC – Call Control

CC sets up, maintains and releases calls, and is used to select the type of service. One of the most important functions of CC is call routing. The Mobile Subscriber ISDN (MSISDN) number, dialled by a caller, includes:

- Country code.
- National destination code identifying the subscriber's operator.
- Code corresponding to the subscriber's HLR.
- The subscriber's "local" number within the HLR.

The call is passed to the GMSC (if the call originates from a fixed network) which knows the HLR corresponding to the MSISDN number. The GMSC asks the HLR for information required to route the call. The HLR in turn interrogates the subscriber's current VLR. The VLR allocates a temporary Mobile Station Roaming Number (MSRN) for the call. The MSRN number is returned by the HLR to the GMSC. Using the MSRN number, the call is routed to subscriber's current MSC/VLR. In the subscriber's current LA, the mobile is paged.

3.5.2 Supplementary Services management

The mobile station and the HLR are the only components of the GSM network involved with Supplementary Services (SS). This makes it easier to introduce new services without major upgrades to the whole network. The various SS available to users are outlined in section 6.3.

3.5.3 SMS (Short Message Services) management

For handling SMS, the GSM network is in contact with a Short Message Service Centre through two interfaces:

+ The **SMS-GMSC,** for Mobile Terminating Short Messages (SMS-MT/PP) sent to the mobile subscriber. It has the same role as the GMSC.
+ The **SMS-IWMSC,** for Mobile Originating Short Messages (SMS-MO/PP) sent by the mobile subscriber.

SMS text messages are sent in the SDCCH signalling channel (See section 5).

3.6 OAM – Operation, Administration and Maintenance

The OAM function is used to monitor, control and reconfigure the system and its various elements. The OSS plays an important part in the OAM function, as do the BSS and NSS.

The BSS and NSS carry out extensive self-testing and monitoring. They pass the information to the OSS, which analyses it as part of ongoing OAM activity in supporting the network.

The BSC performs various OAM functions for the several BTS under its control. Modern BTS have become largely autonomous, able to carry out ever more sophisticated self-maintenance. This greatly reduces system overheads.

At the same time the OMC, the main subsystem within the OSS, has become smarter, further reducing the need for human intervention, for example during periods of peak traffic loads..

3.7 Signalling architecture

The RR, MM and CM network functions generate all the signalling on the air interface to and from the mobile. These GSM signalling messages comprise the top layer of a three-tier signalling network model (Fig 5). The two lower layers are incorporated in the transmission function.

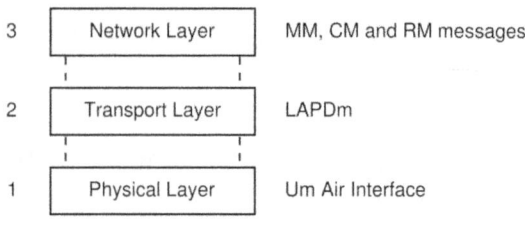

Fig 5 : The signalling network model

3.7.1   Layer 3 protocols

Layer 3 provides a number of different protocols for use by the various network functions in passing messages around the system:

- Radio.
- Mobility.
- Call Control.
- Supplementary Services.
- SMS.
- Test.

3.7.2   Layer 2 – LAPDm

Layer 2 comprises some pre-packaging of the signalling information, allowing it to be reliably transmitted and unpacked at the other end by the physical layer.

The LAPDm layer packages up all the Layer 3 messages in various kinds of *frame* (Note that these are quite different beasts from the TDMA frames we will meet later). These frames are more readily divided than the high-level messages are, into the physical data bursts used at Level 1.

The various Format B frames carry data, whilst the various Format A frames are "fill" frames which pad out the timing gaps in the transmitted bitstream when no actual data is present.

3.7.3   Layer 1

Layer 1 comprises the physical encoding, modulation and transmission of signalling data, corresponding to the Um air interface, and is managed by the Transmission network function.

A key feature of the air interface is the provision of dedicated signalling channels. Several kinds of signalling channel are used, as explained in the next section.

4 The air interface

The air interface connects the mobile stations to the fixed infra-structure. Besides the raw radio transmission standards, it also includes the associated data encoding/recovery, the ways in which the many different kinds of user traffic and GSM signalling data may access the radio link (section 5), and even the signalling messages which pass between base station and mobile to support such access.

GSM uses the Um air interface, which has many sophisticated features to squeeze the last ounce of performance out of the available radio spectrum. Indeed, it pushes things so far out towards the limits that it often fails, and has to rely heavily on a mixture of complicated error-correction along with user tolerance where even that has failed.

Close co-operation between mobile stations and networks from different manufacturers and operators is essential to successful roaming from one network area to another, a key feature of GSM, so the radio interface is correspondingly tightly specified.

The specification also pays great attention to spectrum efficiency, which is affected not just by the nominal capacity of the system, but also by the techniques used to reuse that capacity and to overcome problems of radio propagation, interference and synchronisation.

A variety of techniques are also used to reduce demands on the mobile, simplifying its electronics and extending battery life.

4.1 The GSM radio bands

GSM generally uses two 25 MHz wide bands:

- 890-915 MHz. Uplink (ul), from mobile to base station.
- 935-960 MHz. Downlink (dl), from base station to mobile.

Not all countries use the same GSM frequency band: locally, parts of the GSM band may historically have other uses. America uses a slightly different GSM band around 1900 MHz. A "tri-band" mobile can cope with all three bands, and can be used pretty much world wide.

4.2 Multiple access scheme

The multiple access scheme defines the way in which many calls, to and from different mobile stations within a single cell, can take place simultaneously. The cellular system itself is an example of SDMA – Space Division Multiple Access. Within the cell, GSM primarily uses a mix of frequency and time division multiplexing, with a few extra subtleties.

4.2.1 FDMA

Frequency Division Multiple Access (FDMA) divides the available frequency band into separate narrow-band *carrier frequencies*, or *frequency channels*, which can be allocated to different users. These channels are sometimes referred to as *physical channels*.

The 25 MHz wide GSM band is divided into 124 carrier frequencies each 200kHz wide. Each such frequency channel is identified by its Absolute Radio Frequency Channel Number (ARFCN) in the range 1-124.

A 25 MHz bandwidth should in theory yield 125 carriers, but the lowest 200kHz is left as a guard band to avoid interference between GSM and other services on lower frequencies.

The uplink and downlink bands start 45 MHz apart. Corresponding channels in the ul and dl directions (i.e. having the same ARFCN) will always be separated by the same *duplex spacing* of 45 MHz.

4.2.2 TDMA

Time Division Multiple Access (TDMA) divides the data stream into blocks (Fig 6). Each block is speeded up and transmitted as a short *burst*. More bursts from other data streams are then transmitted, until another burst from the first one is ready. The series of bursts, one from each stream, is called a TDMA *frame*. Once complete, a new frame is started and another burst from each stream is sent. Each data stream occupies the same burst position in successive frames. This burst position is called a *timeslot*.

The data stream is a logical communication channel, or *logical channel* for short. There can be as many logical channels as there are bursts within the frame. In GSM, such a logical channel may carry a user call or signalling messages.

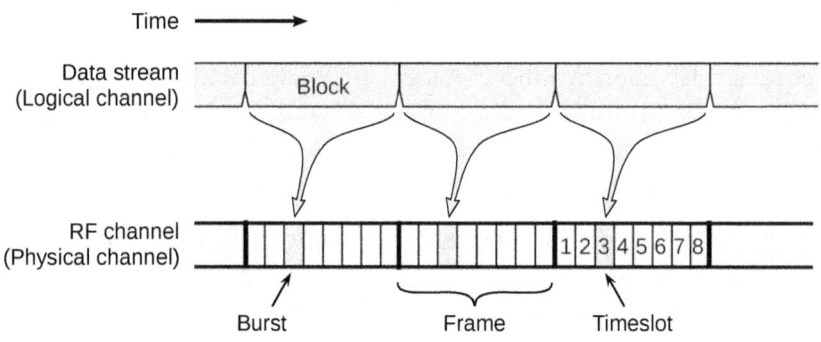

Fig 6 : Time-division multiplexing

GSM uses TDMA within its overall FDMA structure, with each frequency channel being divided in time. A frame lasts 4.615 ms, and is subdivided into 8 bursts of approx. 0.577 ms each.

To simplify the mobile's circuitry it does not transmit on its timeslot while it is receiving from the base station, but transmits three timeslots later; i.e. it waits for three bursts after receiving the base station burst, using the time to reset its circuitry for transmission. This technique is called Time Division Duplex (TDD).

The mobile can only operate on a single frequency at any one time, however traffic and signalling channels may be on different frequencies. These channels must be carefully allocated to different timeslots so that they do not occur simultaneously, giving the mobile time to change frequency during the frame.

4.3 Other radio features

4.3.1 Timing advance

The exact timing of the burst transmissions is critical. Mobiles at different distances from the base station will have differing propagation delays, due to the finite speed of light. Timing advance ensures that signals coming from different mobile stations are synchronised at the base station. The base station monitors the time delay of each mobile. If the bursts from a distant mobile arrive too late and overlap with another's following burst, the base station instructs it to send its bursts a little sooner, i.e. to advance its timing.

The maximum allowable advance sets a theoretical maximum size limit on GSM cells.

4.3.2 Power control

At the same time that the base station performs the timing measurements, it also measures the signal strength from the different mobile stations. It then instructs each mobile to adjust its power level, so that the power received at the base station is similar for each burst.

A base station also controls its own power level. The mobile station monitors the signal from the base station. If the mobile station does not receive a satisfactory signal, it notifies the base station, which then adjusts its own power level on that channel to suit.

4.3.3   Multipath equalisation and the training sequence

A GSM receiver will see not only the direct signal from the trans-mitter, but also spurious signals with differing phases, reflected from buildings, cars, hills, etc., which corrupt the received signal. To overcome this, a prearranged *training sequence* is embedded in the signal. An *equaliser* compares the received and expected training sequences, and calculates the channel transfer function (a measure of the signal distortion). From this it constructs an inverse filter through which the distorted signal is then passed to reverse the distortion and recover the true signal.

4.3.4   Frequency hopping

Radio propagation conditions and associated effects such as multi-path fading vary in time and according to the exact radio frequency. Co-channel interference will also vary in time according to the neighbouring signals and equipment performance. Sometimes, in spite of every trick employed, these will unacceptably degrade a radio channel. By regularly changing everybody's carrier frequencies, any differences in the quality of the individual channels are evened out across the users and reduced to a tolerable level for everybody. This technique is called frequency hopping.

GSM uses a slow frequency hopping method in which each logical channel moves to a different carrier frequency (physical channel) with every TDMA frame, while remaining in the same timeslot.

The sequence of frequencies used for each hop must be synchronised at both ends of the link. A number of algorithms have been developed for this purpose, and the algorithm in use is sent through the Broadcast Control Channels.

Not all channels can be frequency-agile in this way. It is impractical for the common control channels, shared by many mobiles, to hop around.

A base station does not have to use frequency hopping. However, a mobile must be able to accept frequency hopping when the base station is using it.

4.3.5   DTX – Discontinuous transmission

DTX suspends radio transmission during periods of silence, freeing the traffic channel for other users. A person speaks less than 40 percent of the time during a conversation, so DTX significantly reduces demands on the system, allowing extra capacity. DTX employs two related features:

- **Voice Activity Detection (VAD)**, which has to determine whether the audio input represents speech or noise, even if the background noise is loud. If the voice signal is mistakenly treated as noise, the transmitter will cut off producing an unpleasant clipping.
- **Comfort noise**. A problem with DTX is that when there is no voice and the transmitter is off, the receiver hears total silence – the connection appears to be dead. To indicate that the connection is still live, the receiver set creates a low-level "comfort" noise.

4.3.6   Discontinuous reception

Discontinuous reception is used to conserve the mobile station's power. The paging channel is time-multiplexed into subchannels corresponding to individual mobile stations. Each mobile station listens only to its own subchannel and reverts to sleep mode during the time allocated to the other paging subchannels.

4.4 Processing the signal from source to destination

For the purposes of transmission coding, three general types of data can be distinguished:

- User voice, or other audio.
- User digital data.
- GSM signalling data.

All three must pass through the same sequence of process steps when being prepared for radio transmission and recovered at the other end (Fig 7). However the details of each step vary for each type of data: they use different types of logical channel, and the channel to which the data is assigned plays a key role in determining the exact processing path.

4.4.1 Data coding

Speech and other audio is digitally encoded before transmission and then decoded afterwards using a speech codec. The GSM codec is RPE-LTP (Regular Pulse Excitation Long Term Prediction). It analyses the raw audio signal in 20ms blocks, using a model of the human vocal tract and ear to drastically prune unimportant information without significant loss of quality. The output block comprises just 260 bits, giving a data rate of 13 kbps.

User digital data requires only to be divided into blocks of 240 bits, before passing straight to the channel coder.

Signalling messages are encoded into 184 bit Level 2 LAPDm frames, as described in section 3.7.2.

4.4.2 Channel coding

Channel coding reorganises the original information, in order to allow detection and, if possible, correction of the frequent errors which occur during transmission.

The air interface

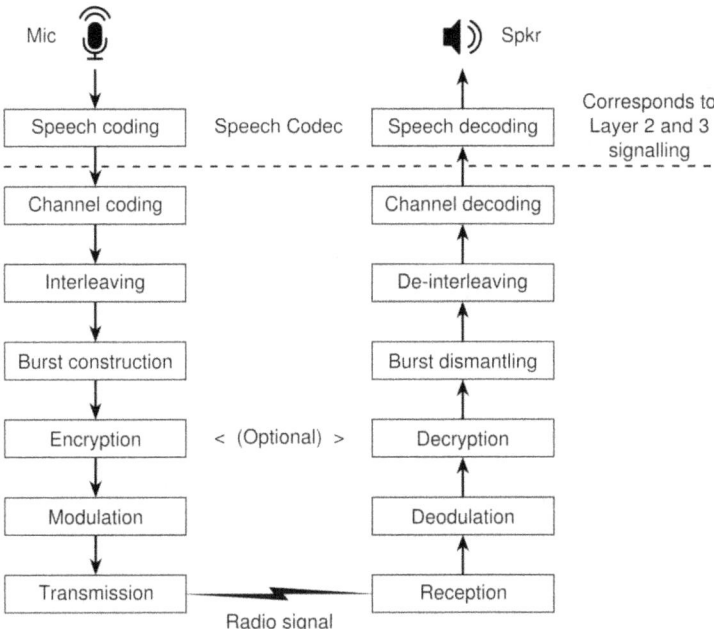

Fig 7 : Processing of speech to radio and back

Signalling data is coded most thoroughly, since a single uncorrected error may be disastrous. User data is coded rather less thoroughly, since it can usually be resent, and audio least thoroughly, since the odd distortion is acceptable. In all cases the basic approach is the same.

The data is first padded out using a *block coding* to prepare it for the next step. The padding may be parity bits (for data redundancy) or just zeroes. *Convolutional coding* is then applied, which has the effect of spreading each slice of data across the block while at the same time adding redundancy. Any transient error within the block thus stands a good chance of being correctable. All the data ends up in blocks of 456 bits.

4.4.3 Interleaving

Interleaving consists in slicing up the data blocks and shuffling the slices, or "leaves," around. This further disperses any section of data across several bursts, decreasing the possibility of losing whole data frames during transmission.

The 456 bit blocks are sliced up and rearranged into 114 bit blocks ready for assembling into bursts. Again, the data types are treated differently, each block being spread across four, eight or twenty-two bursts according to importance. Likewise, each burst will contain data from one, two, or up to six data blocks.

4.4.4 Burst assembling

The burst assembly process adds the guard bits, training sequence, etc. to the 114 bits of a data block, to form the TDMA burst.

4.4.5 Enciphering

Enciphering is intended to protect signalling and user data from eavesdroppers. Data may or may not be enciphered. Not all GSM systems necessarily implement enciphering: different systems may support different levels of enciphering (though the procedures are very similar), and if enciphering is temporarily unavailable (for example if connection to the AuC is lost), services will continue unenciphered in preference to total loss. And some data, such as that exchanged before a mobile is authenticated, cannot be enciphered anyway.

A 64-bit cipher key Kc is computed using the operator's A8 algorithm (stored in the SIM card and the AuC), the subscriber key Ki and the random number RAND generated by the network during authentication (A8 and the authentication algorithm A3 are often combined as a single algorithm which outputs both SRES and Kc, with Kc being stored until needed).

A 114 bit sequence is then calculated using Kc, the A5 algorithm and the 22-bit burst number (T1 + 3 + 2). This bit sequence is XORed with the two 57 bit data blocks of the normal burst.

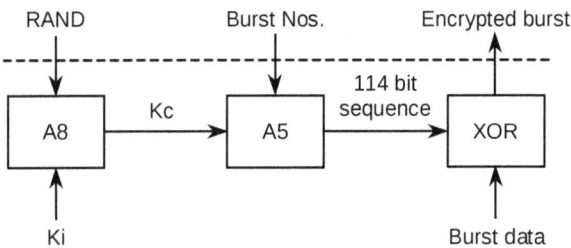

Fig 8 : Enciphering

To decipher the signal, the receiver will use the same A5 algorithm.

The A5 algorithm must support roaming, and so is a GSM standard. Many GSM systems worldwide use the A5/2 algorithm. Networks in COCOM (Western) countries usually support both this and the stronger A5/1 enciphering, which is not universally available.

4.4.6 Modulation

This section will be of little interest to you unless you are a keen RF engineer. Indeed, you will probably understand little of it anyway. Published explanations sometimes get the technical details wrong, but nobody ever seems to notice.

Modulation is the changing of a signal in some characteristic way to convey information. GSM uses Gaussian Minimum Shift Keying (GMSK) for modulation of the RF carrier. This strikes an optimum balance between spectrum efficiency, circuit simplicity and low spurious radiations (reducing interference with adjacent channels).

In shift keying, the RF carrier modulates or shifts between a slightly higher and a lower frequency, with the pattern of changes encoding the data.

Minimum Shift Keying (MSK) makes the difference, or shift, between the two frequencies as small as is theoretically possible to allow reliable detection of each transition.

A Gaussian filter is a bandpass filter which allows the main signal to pass, but rolls off on either side with a certain characteristic to block other frequencies, especially harmonics.

In GMSK the digital signal is first prepared as a stream of sharp transitions between two, high and low, voltages. A Gaussian filter is next applied to the digital signal, to round-off the sharp edges of the digital transitions and in so doing to suppress out-of-band harmonics. This waveform is finally used to modulate the carrier wave between its two MSK frequencies.

5 Channel structures

5.1 General structures

GSM radio transmissions fall into three broad types of data structure:

- User traffic and associated signalling messages.
- Common access control messages.
- Dedicated signalling messages.

Each of these structures has its own variations on the overall scheme of things, using one or more variants of:

- Logical channel types and groupings.
- TDMA frame sequencing.
- TDMA burst structure.

These details are all part of the Um air interface, but are complex enough to deserve their own section.

5.1.1 Logical channels and grouping

A logical channel comprises bursts at a given location within repeating TDMA frames (section 4.2.2). It is allocated to a physical channel and timeslot.

In GSM there are two broad families of logical channel:

- **TCH – Traffic channels:** Carry user speech/audio or digital data information.
- **CCH – Control channels:** Carry signalling messages for network management and channel maintenance. One type also carries user SMS text messages. May be subdivided into common and dedicated control channels.

The control channels are also called signalling channels and Data-mobile channels, Dm. They carry the signalling required to carry out the network management functions. There are four main types of control channel:

- ♦ BCH – Broadcast CHannels.
- ♦ CCCH – Common Control CHannels.
- ♦ DCCH – Dedicated Control CHannels.
- ♦ FACCH – Fast Associated Control CHannels.

Most of these have various sub-types.

Some logical channels only carry sporadic signals: several may share a single timeslot, with each channel being allocated its own pattern of bursts. A small number of channel sub-types may be grouped together to create a particular facility on a given timeslot. Any one group fully occupies its given timeslot. Seven such groupings are defined, numbered I to VII.

5.1.2   TDMA frame sequencing

To aid in synchronising and managing the various logical channels grouped in each timeslot, the TDMA frames (that we met in Section 4.2.2) are strung together into longer *multiframes*:

- ♦ User traffic and associated signalling channels are allocated within a 26 TDMA frame multiframe (26-multiframe). A 5-bit frame counter, T2, cycles from 0 to 25.
- ♦ Common access control and dedicated signalling channels are allocated within a 51-multiframe. A 6-bit frame counter, T3, cycles from 0 to 50.

The 26- and 51- multiframes synchronise (i.e. start together) only once every 1,326 frames. This set is called a *superframe*, and lasts about 6 sec.

The combination of T2 and T3 yields a unique 11-bit *short TDMA frame number* for every frame within the superframe.

Finally, a *hyperframe* lasts for 2,048 superframes, or just under 3½ hrs. An 11-bit superframe counter, T3, cycles from 0 to 2,047.

The combination of T1, T2 and T3 yields a unique 22-bit *TDMA frame number* for every frame within the hyperframe.

5.1.3 Burst structure

The burst is the basic unit of time in a TDMA system. Data is transmitted at 3.69 µs per bit. The *normal burst* is by far the most common type, and may carry speech or data information. Three other types of burst are used by certain common access control channels:

* The frequency-correction burst, used by the FCCH.
* The synchronisation burst, used by the SCH.
* The random access burst, used by the RACH.

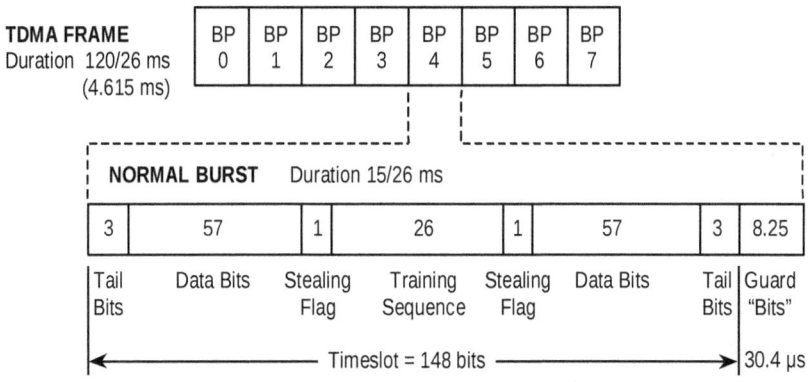

Fig 9 : The TDMA frame and normal burst

The normal burst (Fig 9) lasts approximately 0.577ms and has a length of 156.25 bits:

> The actual data within the timeslot is divided into two groups of 57 bits each.

Two stealing flags (S) indicate to the receiver whether the burst is carrying traffic (TCH) or signalling data (FACCH).

The training sequence has a length of 26 bits. The receiver will use it to remove any distortions due to multipath propagation, as described elsewhere.

Three zero-value tail bits (T), at the start and the end of the burst, cover the periods of ramping up and down of the mobile's power. A guard period (GP) at the end of 30.4 μs avoids an unsynchronised mobile overlapping another outside the RF ramping time.

5.2 User traffic and associated signalling

Channel groups I to III provide for the various kinds of user traffic and associated signalling. A typical group comprises a given type of TCH (or, if required, FACCH) together with the associated SACCH.

These channels are frequency-agile, with the group hopping together as a whole.

5.2.1 TCH – Traffic Channels

There are various kinds of traffic channel. Full-rate channels have twice the data rate of half-rate channels:

♦ **TCH/F** – Traffic CHannel Full-rate, also called Bearer-mobile channel, Bm.

♦ **TCH/H** – Traffic CHannel Half-rate, also called Low-mobile channel, Lm.

Different sub-types carry user speech/audio or digital data. These yield seven combinations:

♦ TCH/FS – Traffic CHannel Full-rate Speech (Group I).

♦ TCH/HS – Traffic CHannel Half-rate Speech (Group II).

- TCH/F9.6, 4.8 and 2.4 – Traffic CHannel Full-rate, data at 9.6, 4.8 or 2.4 kBaud.
- TCH/H4.8 and 2.4 – Traffic CHannel Half-rate, data at 4.8 or 2.4 kBaud.

5.2.2   SACCH – Slow Associated Control Channels

SACCH are a type of DCCH used for channel control and main-tenance. Every TCH has an associated SACCH. SACCH are also associated with signalling channels, and more about them is explained in that subsection.

5.2.3   FACCH – Fast Associated Control Channels

The FACCH replace all or part of a traffic channel when needed, e.g. during handover, to transmit urgent SDCCH signalling information.

The FACCH uses the same Normal burst as the TCH, but sets the stealing flags to 1, to signify that it has stolen the burst from a TCH.

5.2.4   The 26-frame multiframe

Traffic and the associated signalling are organised using a 26-multi-frame. The multiframe lasts 120 ms. Within the multiframe, downlink and uplink traffic channels are separated by 3 bursts.

Different types of traffic channel employ 26-multiframes with subtly different internal structures. For example the full-rate traffic channel (Fig 10) has:

24 frames reserved for the call traffic (TCH, or FACCH as required).

1 frame for the Slow Associated Control Channel (SACCH).

The last frame is unused. This idle frame gives the mobile station time to perform other functions, such as measuring the signal strength of neighbouring cells.

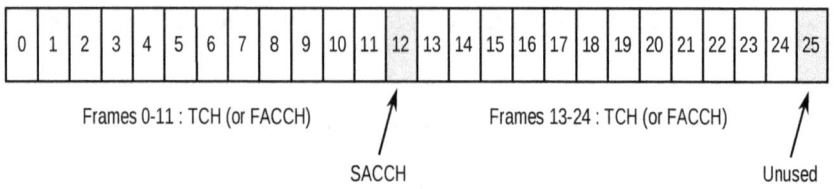

26-FRAME MULTIFRAME   Duration 120 ms

| 0 | 1 | 2 | 3 | 4 | 5 | 6 | 7 | 8 | 9 | 10 | 11 | 12 | 13 | 14 | 15 | 16 | 17 | 18 | 19 | 20 | 21 | 22 | 23 | 24 | 25 |

Frames 0-11 : TCH (or FACCH)   Frames 13-24 : TCH (or FACCH)

SACCH   Unused

Fig 10 :   Type III 26-multiframe

5.3 Common access control

Common access control channels are found in Group types IV, V and VI. They cannot frequency hop.

Every base station has a *beacon frequency*, on which it broadcasts basic access information at full power. The beacon carries two families of access control channel:

- **BCH** – Broadcast Channels
- **CCCH** – Common Control Channels

The basic channel group for the beacon is Type IV. In a quiet cell, the beacon may also carry dedicated signalling channels (Type V). In a busy cell, further BCCH and CCCH may be provided as one or more Type VI groups on additional timeslots, and possibly on other frequencies.

5.3.1   BCH – Broadcast Channels

The Broadcast channels (BCH) are broadcast by the base station to provide the mobiles with the information they need to synchronise with the network. There are three different types of BCH:

- FCCH – Frequency-Correction Channel. Carries the frequency reference of the system necessary to synchronise the mobile station with the network. It uses a special frequency-correction

44

burst, which has the same length as the normal burst and the same tail and guard at each end, but in between is the frequency correction sequence, comprising all zeroes. The FCCH has a distinctive radio signature by which the beacon frequency may be recognised.

♦ SCH – Synchronisation Channel. Carries the training sequence or key necessary to recover the BCCH signal. It uses a special synchronisation burst, which has the same length as the normal burst but a different structure. Information on the SCH includes: LAC – Network Location Area Code.

SSIC – Base Station Identifier Code, comprising:

BCC – Base Station Colour Code (3 bits),

NCC – Network (PLMN) Colour Code (3 bits).

Short TDMA frame number (T2 + T3).

♦ BCCH – Broadcast Control Channel. Carries the parameters necessary to identify and access the network. The main BCCH is always found on Timeslot 0. This is the station's *base channel*. Information on the BCCH includes: LAC – Network Location Area Code.

MNC – identifies the operator.

Cell ID parameters, some of which are specific to the operator.

Cell channel information (not necessarily for all physical channels).

Cell options available.

Common channel configuration information.

PAGCH details.

RACH control parameters, including the initial power level for mobiles.

5.3.2   CCCH – Common Control Channels

Once the mobile station has accessed the base channel, the CCCH will help to establish calls between the mobile and the network. There are three types:

- RACH – Random Access Channel, on which the mobile initially requests access to the network. The timing advance has not yet been set, so the mobile uses a special random access burst which is shorter than normal, to ensure that it will fit in between the properly-timed bursts of other mobiles. The RACH is the only uplink access control channel, all the others are downlink-only.

- AGCH – Access Grant Channel, on which the base station responds to the mobile's initial request for access (see below) by telling the mobile:
  Channel allocation details.
  The timing advance.

- PCH – Paging Channel, on which the base station advises the mobile of an incoming call.

In practice, one or more PCH and (when required) the AGCH may be combined into a single PAGCH.

5.4 Dedicated signalling

Dedicated signalling channels may be grouped with common access channels (Type V) or in their own grouping (Type VII).

5.4.1   DCCH – Dedicated Control Channels

The DCCH are used for message exchange to/from mobiles. There are three types:

- SDCCH – Standalone Dedicated Control Channel, for the bidirectional exchange of signalling information between mobile and BS. Also carries user SMS text messages.

- SACCH – Slow Associated Control Channel, for channel control and maintenance. SACCH are always used in association with a TCH (for traffic) or an SDCCH (for fast signalling). Messages may be exchanged between mobiles, as well as to/from the BS.
- CBCH – Cell Broadcast CHannel, for SMS broadcasts to multiple mobiles.

6 GSM services

The GSM system provides many services beyond basic telephony, which have been introduced progressively over time. Some of these are mandatory GSM standards, others may or may not be implemented by the network operator (optional services). There are three main types of service:

- Teleservices.
- Bearer services.
- Supplementary Services.

Each of these services is likely to have its own particular signalling protocol, of greater or lesser complexity, overlaid on the basic GSM signalling system.

6.1 Teleservices

A teleservice provides a link between two users for the transport of conversations and messages.

Telephony. The basic voice call service of any telephone network.

Emergency calls. A special high-priority telephony service. Emergency calls can be made from an ME even if there is no SIM card present and therefore no identifiable subscriber account.

Voice mail. This service corresponds to an answering machine.

Telefax and Fax mail. Provides Group 3 Fax. Mail allows transmission to a fax machine which is currently offline.

Short Message Services. An SMS message comprises a string of up to 160 text characters. Messages can be sent to or from a mobile. If the destination mobile is powered off, the message is stored. The SMS Cell Broadcast (SMS-CB) service will send a message of up to 93 characters to all mobiles in a particular geographical area. Many more optional service features exist.

6.2 Bearer services

A bearer service transports digital user data. Like all GSM connections it is is circuit-switched directly between the caller and the recipient. Because of this these services are unsuited to packet-switched digital networks such as the Internet. While a connection to a packet-switched network can be made via a modem and circuit-switched call to a packet-switching Point of Presence (PoP), the data rate obtainable is severely limited and this led to the development of GPRS or "2.5G" mobile, which is not discussed further.

Some bearer services are:

Asynchronous and synchronous data, 300-9600 bps.

Alternate speech and data, 300-9600 bps.

Asynchronous PAD (packet assembler/disassembler) access, 300-9600 bps.

Synchronous dedicated packet data access, 2400-9600 bps.

6.3 Supplementary services

Supplementary services provide various facilities for handling calls, and are not directly concerned with the call content.

AoC – Advice of Charge. Provides the user with call charge information.

CLIP – Calling Line Identity Presentation. (Optional service). Informs the called user of the calling user's ISDN.

CLIR – Calling Line Identity Restriction. (Optional service). The calling user may withhold information, e.g. their ISDN, from the called user.

Call hold. Puts the active call on hold, for example to stop a conversation with someone in the same room from being picked up.

CW – Call Waiting. Alerts the user, during a call, to a second incoming call. The user can answer, reject or ignore the incoming call.

Multiparty communication. Conference calls between multiple users.

Call Forwarding. The user can forward incoming calls to another number; if the called mobile is busy (CFB) or unreachable (CFNRc), if there is no reply (CFNRy), or unconditionally (CFU).

CoLP – Connected Line identity Presentation. (Optional service). Where a call is forwarded, informs the calling user of the ISDN to which the call is to be forwarded.

CoLR – Connected Line identity Restriction. (Optional service). Where a call is forwarded, the called user may withhold information, e.g. their ISDN, from the calling user.

Call Barring. The subscriber can bar the user from making or receiving certain calls:

- BAOC – Barring of All Outgoing Calls.
- BOIC – Barring of Outgoing International Calls.
- BOIC-exHC – Barring of Outgoing International Calls except those directed toward the Home PLMN Country.
- BAIC – Barring of All Incoming Calls.
- Barring of incoming calls when roaming (optional service).

Operator determined barring. (Optional service). Restriction of different services and call types by the operator.

CUG – Closed User Group. (Optional service). Calls may only be made to or from other group members and certain numbers).

List of acronyms

A3	Authentication algorithm
A5	Ciphering algorithm
A8	Ciphering key computation
AGCH	Access Grant CHannel
AMPS	Advanced Mobile Phone Service
AoC	Advice of Charge
ARQ	Automatic Repeat reQuest mechanism
ARFCN	Absolute Radio Frequency Channel Number
AUC	Authentication Centre
BAIC	Barring of All Incoming Calls
BAOC	Barring of All Outgoing Calls
Baud	Unit of digital transfer rate: 1 Baud = 1 bit per second
BOIC	Barring of Outgoing International Calls
BOIC-exHC	Barring of Outgoing International Calls – except those directed toward the Home PLMN Country.
BCCH	Broadcast Control CHannel
BCH	Broadcast CHannel
BER	Bit Error Rate
bps	bits per second
BSC	Base Station Controller
BSS	Base Station Subsystem
BTS	Base Transceiver Station
CC	Call Control
CCCH	Common Control CHannel
CFB	Call Forwarding on mobile subscriber Busy
CFNRc	Call Forwarding on mobile subscriber Not Reachable
CFNRy	Call Forwarding on No Reply
CFU	Call Forwarding Unconditional
CGI	Cell Global Identity
C/I	Carrier-to-Interference ratio

CLIP	Calling Line Identification Presentation
CLIR	Calling Line Identification Restriction
CM	Communication Management
CoLP	Connected Line identification Presentation
CoLR	Connected Line identification Restriction
CUG	Closed User Group
CW	Call Waiting
DCS	Digital Cellular System
DCCH	Dedicated Control CHannel
DTX	Discontinuous Transmission
EDGE	Enhanced Data rates for GSM Evolution. An enhancement to GPRS.
EIR	Equipment Identity Register
FACCH	Fast Associated Control CHannel
FCCH	Frequency-Correction CHannel
FDMA	Frequency Division Multiple Access
FEC	Forward Error Correction code
FER	Frame Erasure Rate
GIWU	GSM InterWorking Unit
GMSC	GSM Mobile services Switching Centre
GMSK	Gaussian Minimum Shift Keying
GP	Guard Period
GPRS	General Packet Radio Services. 2.5G packet-switching extension to GSM.
GSM	Global System for Mobile communications
HLR	Home Location Register
HSCSD	High Speed Circuit Switched Data. Enhancement to GSM.
Hz	Frequency in Hertz. 1 Hz = 1 cycle per second.

List of acronyms

IMEI	International Mobile Equipment Identity
IMSI	International Mobile Subscriber Identity
ISDN	Integrated Services Digital Network
LA	Location Area
LAI	Location Area Identity
LAPDm	Link Access Protocol Data mobile
LOS	Line-Of-Sight
MM	Mobility Management
MS	Mobile Station
MSC	Mobile services Switching Centre
MSISDN	Mobile Station ISDN number
MSRN	Mobile Station Roaming Number
NSS	Network and Switching Subsystem
OAM	Operation, Administration and Maintenance
OMC	Operation and Maintenance Centre
OSS	Operation and Support Subsystem
PAD	Packet Assembler Disassembler
PCH	Paging CHannel
PCS	Personal Communications Services
PDC	Personal Digital Cellular
PIN	Personal Identification Number
PLMN	Public Land Mobile Network
PSPDN	Packet Switched Public Data Network
PSTN	Public Switched Telephone Network
RACH	Random Access CHannel
RF	Radio Frequency
RPE-LTP	Regular Pulse Excitation Long Term Prediction
RR	Radio Resources management

S	Stealing flags
SACCH	Slow Associated Control CHannel
SCH	Synchronisation CHannel
SDCCH	Standalone Dedicated Control CHannel
SIM	Subscriber Identity Module
SMS	Short Message Services
SMS-CB	Short Message Services Cell Broadcast
SMS-MO/PP	Short Message Services Mobile Originating/Point-to-Point
SMS-MT/PP	Short Message Services Mobile Terminating/Point-to-Point
SNR	Signal to Noise Ratio
SRES	Signed RESult
SS	Supplementary Services
T	Tail bits
TACS	Total Access Communication System
TCH	Traffic CHannel
TCH/F	Traffic CHannel/Full rate
TCH/H	Traffic CHannel/Half rate
TDMA	Time Division Multiple Access
TMSI	Temporary Mobile Subscriber Identity
VAD	Voice Activity Detection
VLR	Visitor Location Register

www.ingramcontent.com/pod-product-compliance
Lightning Source LLC
Chambersburg PA
CBHW061222180526
45170CB00003B/1117